discovermore
The Human Body

How Our Respiratory System Works

Dwight Morris

Britannica
Educational Publishing
IN ASSOCIATION WITH
THE ROSEN PUBLISHING GROUP

Published in 2026 by Britannica Educational Publishing (a trademark of Encyclopædia Britannica, Inc.) in association with The Rosen Publishing Group, Inc.
2544 Clinton Street, Buffalo, NY 14224

Distributed exclusively by Rosen Publishing.
To see additional Britannica Educational Publishing titles, go to rosenpublishing.com.

Portions of this work were originally authored by Christine Figorito and published as *The Lungs in Your Body*. All new material in this edition was authored by Dwight Morris.

Editor: Greg Roza
Book Design: Rachel Rising

Photo Credits: Cover, (series background) Dai Yim/Shutterstock.com; Cover, p. 5 Magic3D/Shutterstock.com; p. 4 H-studio/Shutterstock.com; p. 7 Designua/Shutterstock.com; pp. 7, 11, 19 Magic mine/Shutterstock.com; p. 8 Inna Kharlamova/Shutterstock.com; p. 9 eveleen/Shutterstock.com; p. 10 Khorzhevska/Shutterstock.com; p. 12 As Ashik17/Shutterstock.com; p. 13 1000 Words/Shutterstock.com; p. 14 Daisy Daisy/Shutterstock.com; p. 15 https://commons.wikimedia.org/wiki/File:Bronchiolar_epithelium_3_-_SEM.jpg; p. 17 Tatsiana Matusevich/Shutterstock.com; p. 17 Nerthuz/Shutterstock.com; p. 19 Antonio Guillem/Shutterstock.com; pp. 21, 22 BlueRingMedia/Shutterstock.com; p. 21 Tatsiana Matusevich/Shutterstock.com; p. 23 Robsonphoto/Shutterstock.com; p. 24 Lopolo/Shutterstock.com; p. 25 Ekaterina Vidiasova/Shutterstock.com; p. 26 Gorgev/Shutterstock.com; p. 27 FOTOGRIN/Shutterstock.com; p. 28 PeopleImages.com - Yuri A/Shutterstock.com; p. 29 VH-studio/Shutterstock.com.

Cataloging-in-Publication Data
Names: Morris, Dwight.
Title: How our respiratory system works / Dwight Morris.
Description: New York : Britannica Educational Publishing, in association with Rosen Educational Services, 2026. | Series: Discover more: the human body | Includes glossary and index.
Identifiers: ISBN 9781641905060 (library bound) | ISBN 9781641905053 (pbk) | ISBN 9781641905077 (ebook)
Subjects: LCSH: Respiratory organs--Juvenile literature. | Human physiology--Juvenile literature.
Classification: LCC QP121.M677 2026 | DDC 611'.2--dc23

Manufactured in the United States of America

CPSIA Compliance Information: Batch #CSBRIT26. For further information contact Rosen Publishing at 1-800-237-9932.

Contents

Always Breathing 4
About the Lungs 6
Working Together 8
Lungs at Work 10
Sound Advice 12
Fighting Illness 14
Lung Structure 16
Breathe Deep 20
Let It Out 22
Illnesses of the Lungs 24
Smoking and Vaping 26
Healthy Lungs, Healthy Body 28
Glossary 30
For More Information 31
Index 32

Always Breathing

Even when you sit perfectly still, your chest rises and falls. That is because your lungs are working to keep you breathing. Breathing is **involuntary**. Your lungs do it without you thinking about it.

Lungs are the main organs, or body parts, of the respiratory system. All animals that have a backbone and breathe air (like you!) have lungs. When an animal inhales (breathes in), air filled with oxygen flows into the lungs. The body needs a steady supply of oxygen to live. When an animal exhales (breathes out), a different gas flows out of the lungs.

You don't need to think about breathing. It just happens!

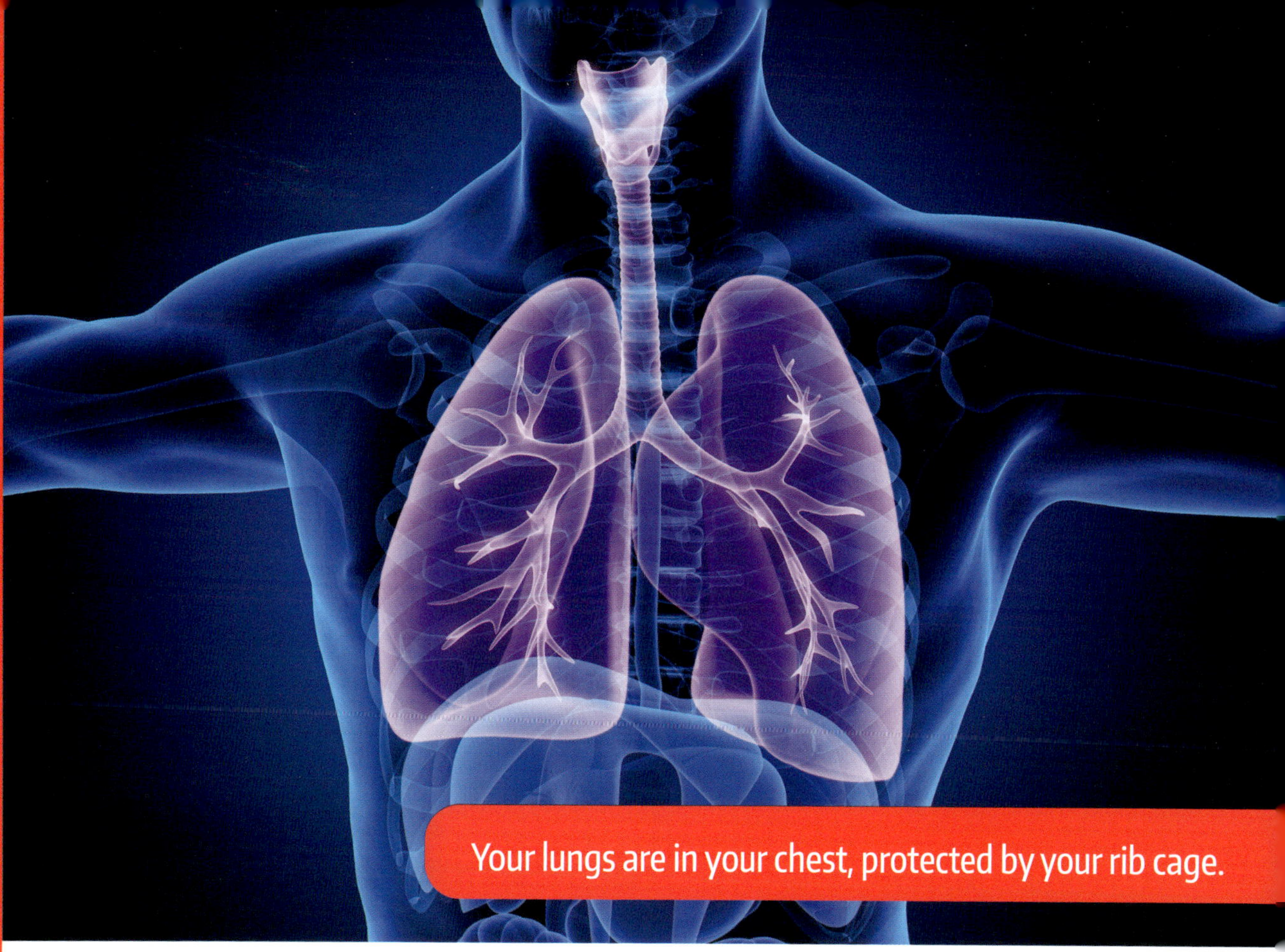

Your lungs are in your chest, protected by your rib cage.

Other organs in the body help with breathing. The brain, for example, is the organ that makes sure the lungs keep working all the time, even when you are asleep.

WORD WISE

SOMETHING THAT IS INVOLUNTARY IS AN ACTION THAT HAPPENS WITHOUT US THINKING ABOUT IT, SUCH AS BREATHING. OUR HEART ALSO WORKS INVOLUNTARILY TO MOVE BLOOD AROUND OUR BODY.

About the Lungs

Everyone has two lungs, one on each side of the chest. Together, they make one of the largest organs in the body. Each lung has about 300 million tiny air sacs. If it were possible to stretch these air sacs out and lay them flat, your lungs would cover an area as large as a tennis court!

The left lung is a little smaller than the right lung. Its smaller size makes room for the heart, which sits between them. The lungs are protected by the rib cage, which surrounds them. The lungs sit on a muscle called a diaphragm. In normal breathing the diaphragm and muscles between the ribs automatically tighten and relax, allowing the lungs to function.

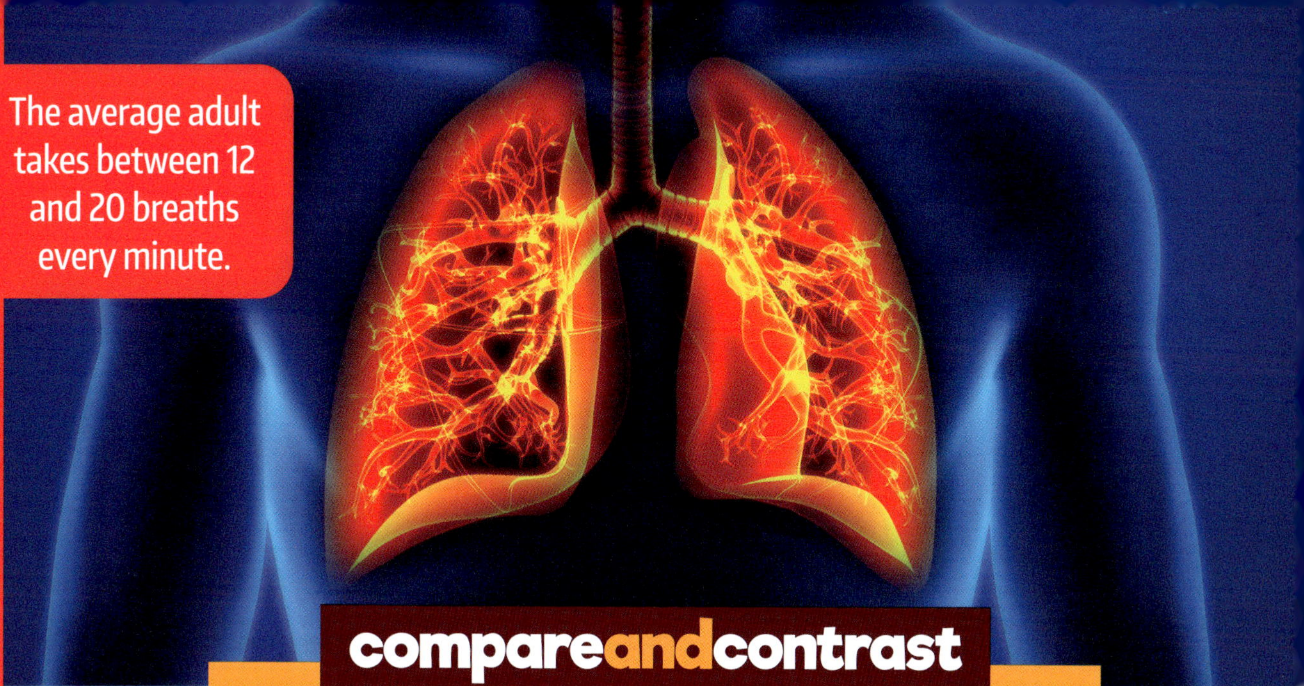

The average adult takes between 12 and 20 breaths every minute.

compareandcontrast

What parts of the body help the lungs do their job? How?

The diaphragm functions involuntarily. When it moves down, the lungs draw in air. When it moves up, the lungs release air.

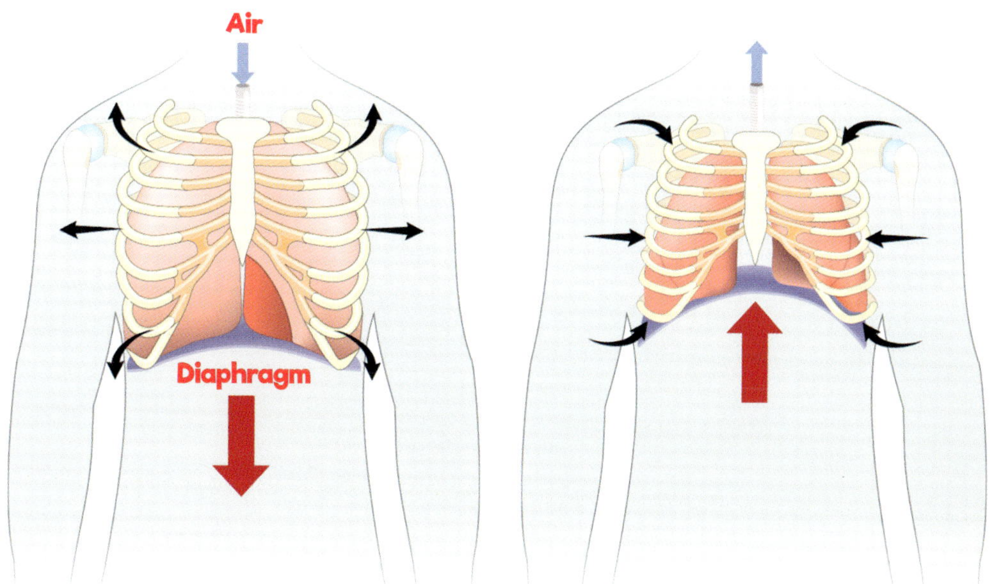

Working Together

A body's breathing relies on its entire respiratory system. Lungs do not breathe by themselves. The human respiratory system includes a number of parts: mouth, nose, throat, voice box, windpipe, bronchial tubes, lungs, and diaphragm.

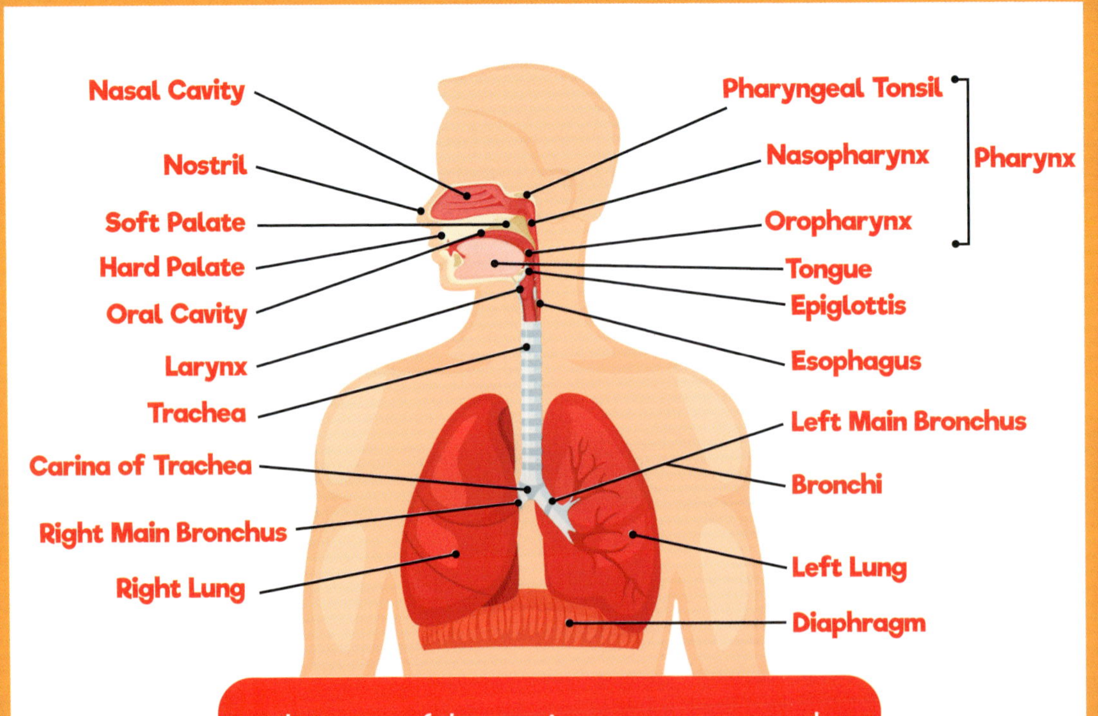

The parts of the respiratory system work together to allows us to breathe.

The voice box is also called the larynx. The windpipe is called the trachea.

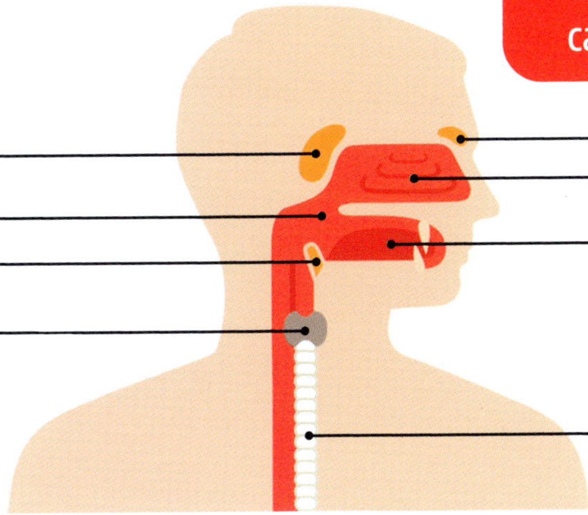

Air enters the respiratory system through the nose or mouth. The throat has two passageways: one for air and one for food. The passage for food is called the esophagus, which is part of the digestive system. The upper part of the air passage contains the voice box. The lower part contains the windpipe. The windpipe separates into two main branches called bronchial tubes, or bronchi. Each bronchi leads to a different lung.

Consider This

The windpipe and the esophagus are very close together. What happens when food or water enters the windpipe? What happens when air enters the esophagus?

Lungs at Work

The lungs provide the body with oxygen, which is a gas that the body needs in order to survive. The lungs also allow the body to get rid of another gas called carbon dioxide.

Plants and people work together! People breathe out carbon dioxide, which plants need to make food during photosynthesis. Oxygen is a waste product of photosynthesis, which people need to live.

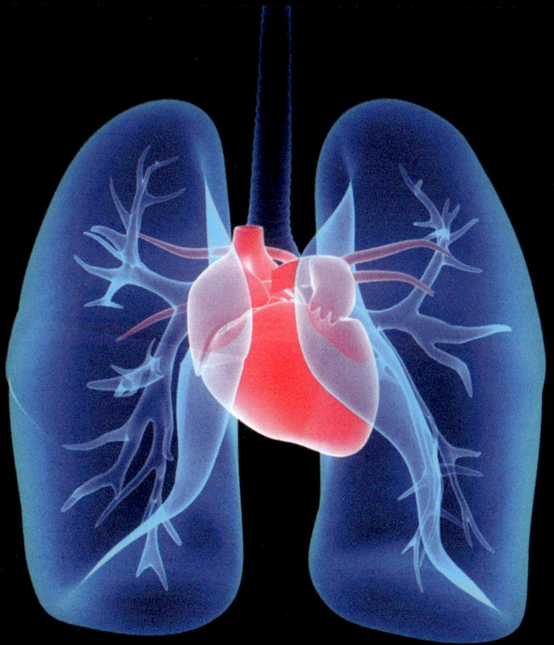

The lungs and the heart work together to take in oxygen and send it to all areas of the body. They are both part of the cardiopulmonary system.

When you breathe in, air enters the lungs. Tiny blood vessels within the lungs **absorb** oxygen from the air. The blood then carries the oxygen through other blood vessels to the heart, and then to all the cells of the body. Cells are basic units that make up all living things. The cells take in the oxygen and release carbon dioxide back into the blood. The blood brings the carbon dioxide to the lungs. Then the lung release the carbon dioxide when you exhale.

WORD WISE

ABSORB MEANS TO TAKE IN. FOR EXAMPLE, A PAPER TOWEL ABSORBS (SOAKS UP) WATER.

Sound Advice

Lungs allow us to talk. Your voice box sits at the top of the windpipe. Inside the voice box are vocal cords. These are a pair of stretchy folds that create the voice. As air exhales from your lungs, it passes between the vocal cords and makes them **vibrate**. This vibration makes sounds. Your voice box makes these sounds louder. Then they pass through your throat, mouth, and nose.

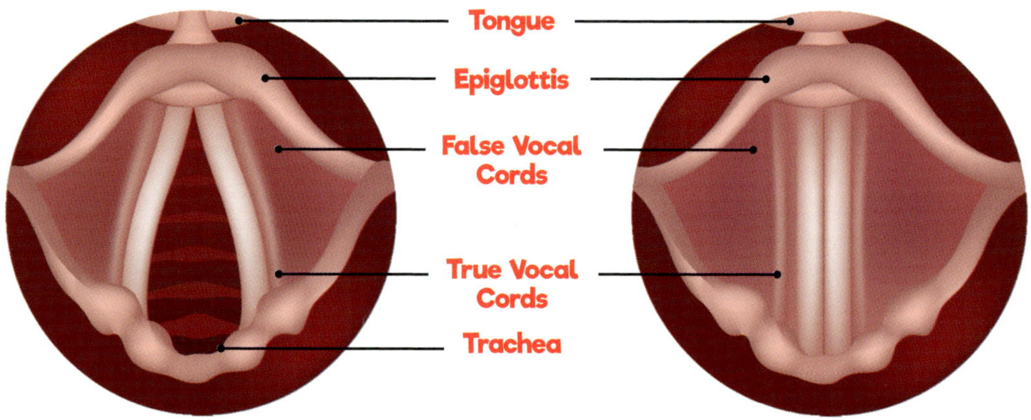

Much of the voice box is made of soft bone called cartilage and other soft tissues that are attached to each other and to surrounding structures.

Our vocal cords, lungs, and mouth work together to allow us to talk, sing, and shout.

You use your tongue, teeth, lips, jaw, and palate (roof of the mouth) to shape the sounds to make words. If you want to say many words quickly or speak loudly, you need to exhale more air. To yell, you need even more air from your lungs.

WORD WISE

VIBRATE MEANS TO MOVE BACK AND FORTH QUICKLY. GUITAR STRINGS VIBRATE WHEN PLUCKED AND MAKE MUSIC.

Fighting Illness

The lungs help keep harmful things, such as germs, out of the body. Inside the lungs are cells that produce mucus. Mucus is a sticky substance that traps pollutants and **bacteria**. Tiny hairs called cilia also keep dirt and bacteria out of the lungs. They move in waves to push the mucus up out of the airways.

Coughing is your body's way of moving mucus, dust, and germs out of the lungs.

Cilia move back and forth to move dirt away from the lungs.

When you cough, the mucus as well as the bacteria and other irritating particles are forced out of the body. The lungs also have cells that find foreign particles and cover them up or destroy them.

WORD WISE

BACTERIA ARE SINGLE-CELLED ORGANISMS FOUND ALMOST EVERYWHERE IN THE WORLD. SOME BACTERIA CAN CAUSE ILLNESS. SOME ARE GOOD FOR OUR HEALTH.

Lung Structure

The two lungs are not exactly the same size. The right lung has three rounded sections called lobes. The left lung has two lobes. The smaller left lung makes room in the chest for the heart.

Tubes used for carrying air connect to the lungs in a branchlike manner. The series begins with the windpipe. Before it enters the lungs, it branches into the two tubes called bronchi (singular: bronchus). Each of these leads to a different lung. The bronchial tubes then divide into a web of smaller tubes called bronchioles. There are about 30,000 bronchioles in just one lung.

compare and contrast

The many branches of the lungs look somewhat like the roots or branches of a tree. How are bronchioles the same as or different from tree roots and branches?

Healthy lungs are pink and spongy. Lungs that are unhealthy may look gray or black and smaller than healthy lungs.

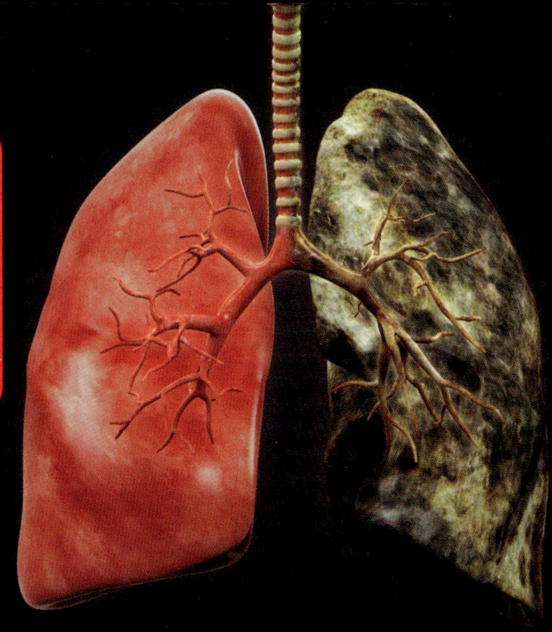

Bronchioles are the smallest branches in the lungs. They are less than a millimeter wide.

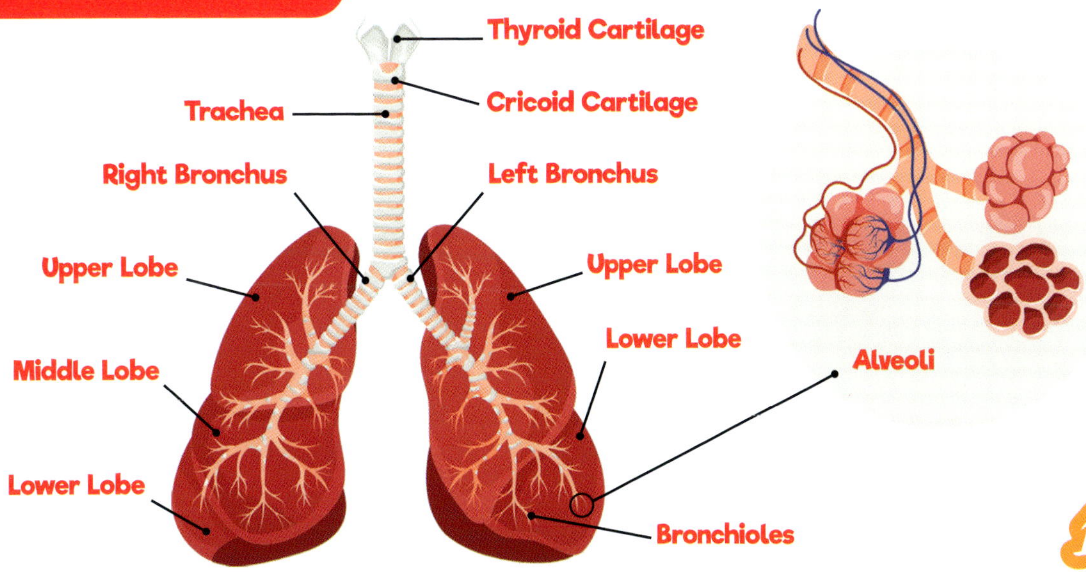

Each bronchiole leads to many tiny air sacs called alveoli. They are what make the lungs feel spongy. They are covered with blood vessels called capillaries. These capillaries are so small that blood cells must line up one by one to get through them! Although tiny, alveoli and capillaries are important. Through them, oxygen gets to the blood cells. The blood cells then deliver the oxygen to your entire body.

compare and contrast

How are alveoli and capillaries the same? How are they different?

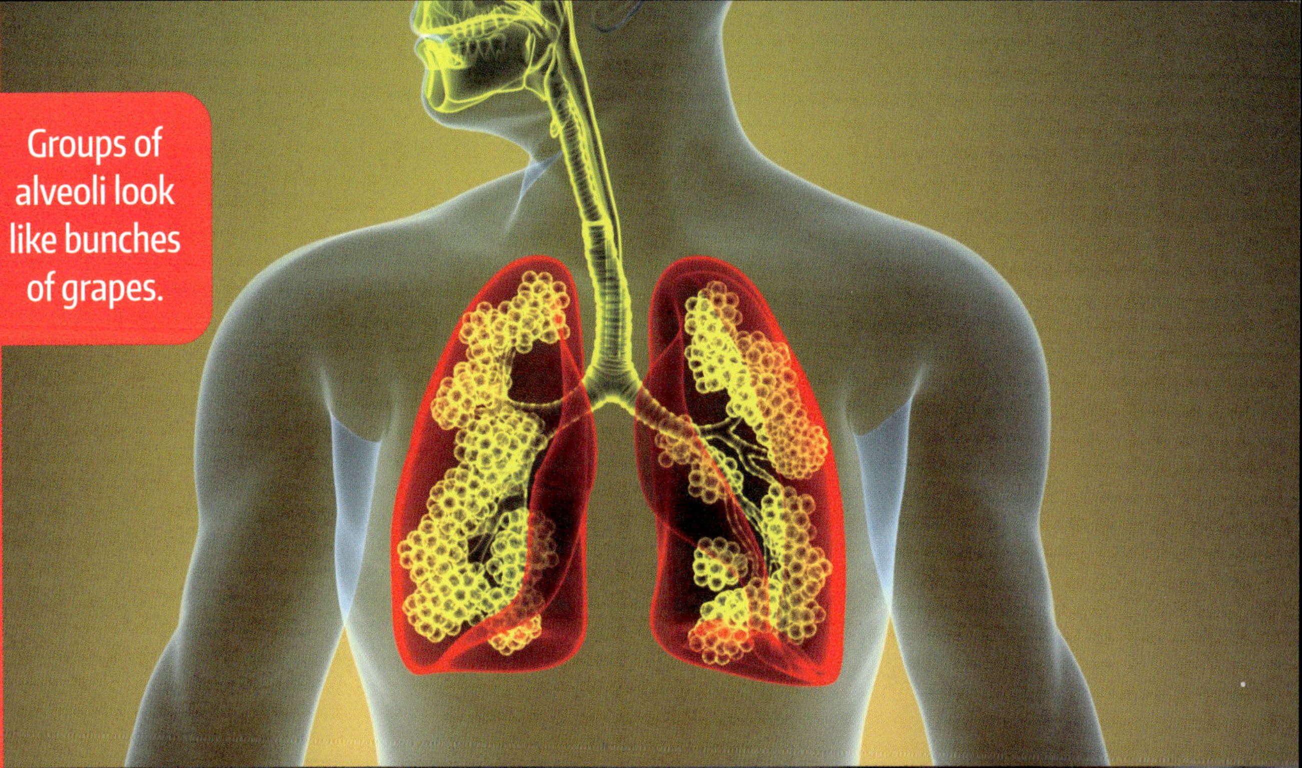

Groups of alveoli look like bunches of grapes.

When you are scared, your heart rate increases and you breathe faster than normal. This helps get oxygen to your cells quickly, allowing your body to respond to danger.

Breathe Deep

When inhaling, or breathing in, the muscle at the base of the lungs, called the diaphragm, moves down and the ribs move out. The area inside your chest grows larger. Air then flows in through your mouth or nose, down the windpipe, and into your lungs.

Oxygen gas in the air is able to pass through the thin walls of the alveoli into the capillaries. Red blood cells inside the capillaries pick up the oxygen and take it to the heart. Then the heart sends the oxygenated blood throughout the body.

The diaphragm contracts, or tightens and becomes smaller, when we breathe in.

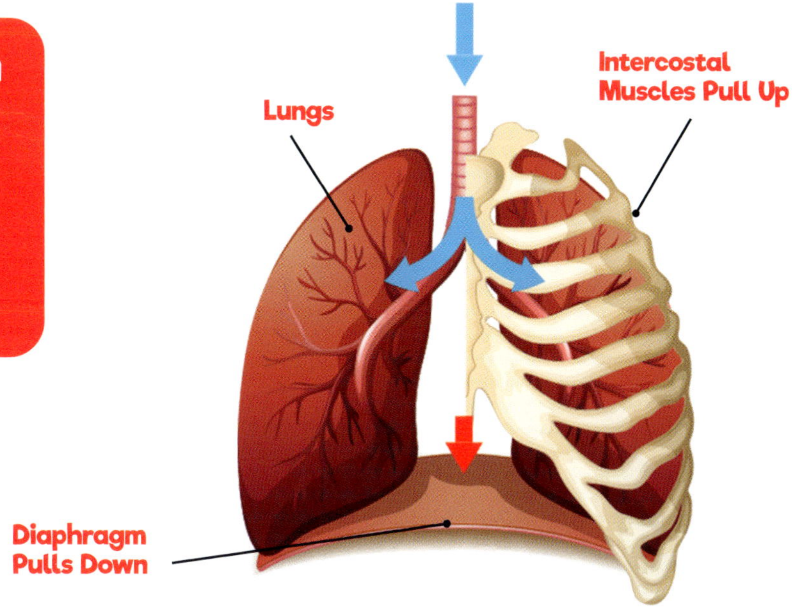

compareandcontrast

Why might you take short, quick breaths rather than long, deep ones?

There are between 400 million to 700 million alveoli in a set of adult lungs.

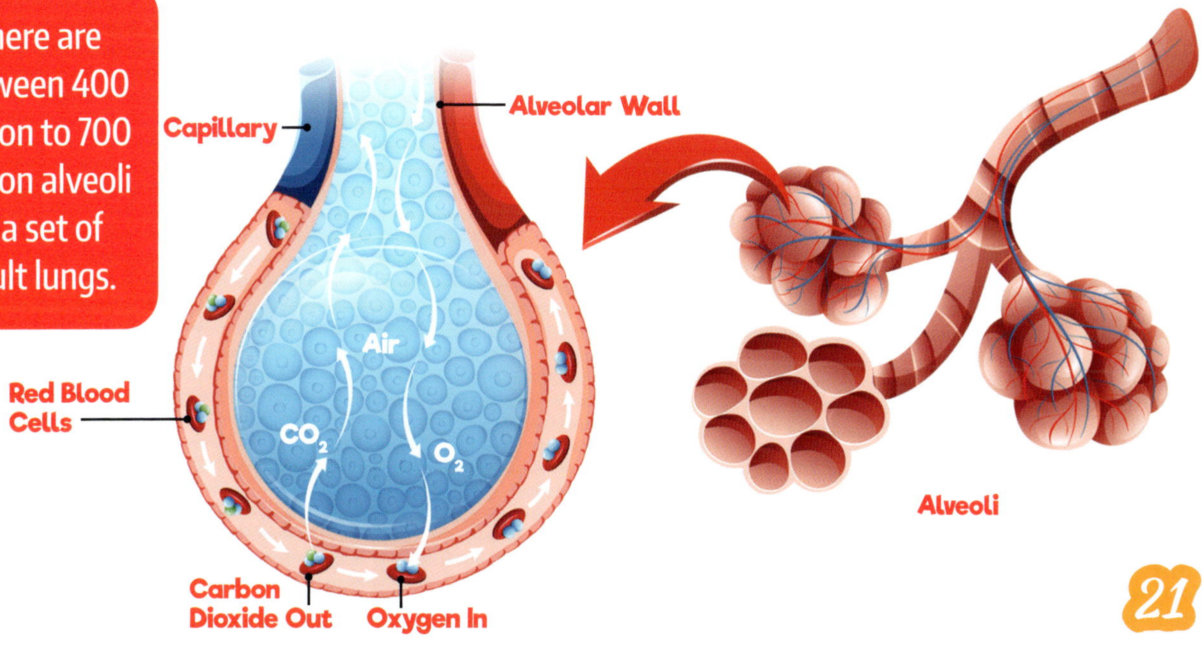

Let It Out

Exhaling, or breathing out, is important for ridding the body of waste gas. As red blood cells deliver oxygen to the body's cells, they pick up waste gas that the cells do not need. These waste gases include carbon dioxide and water vapor.

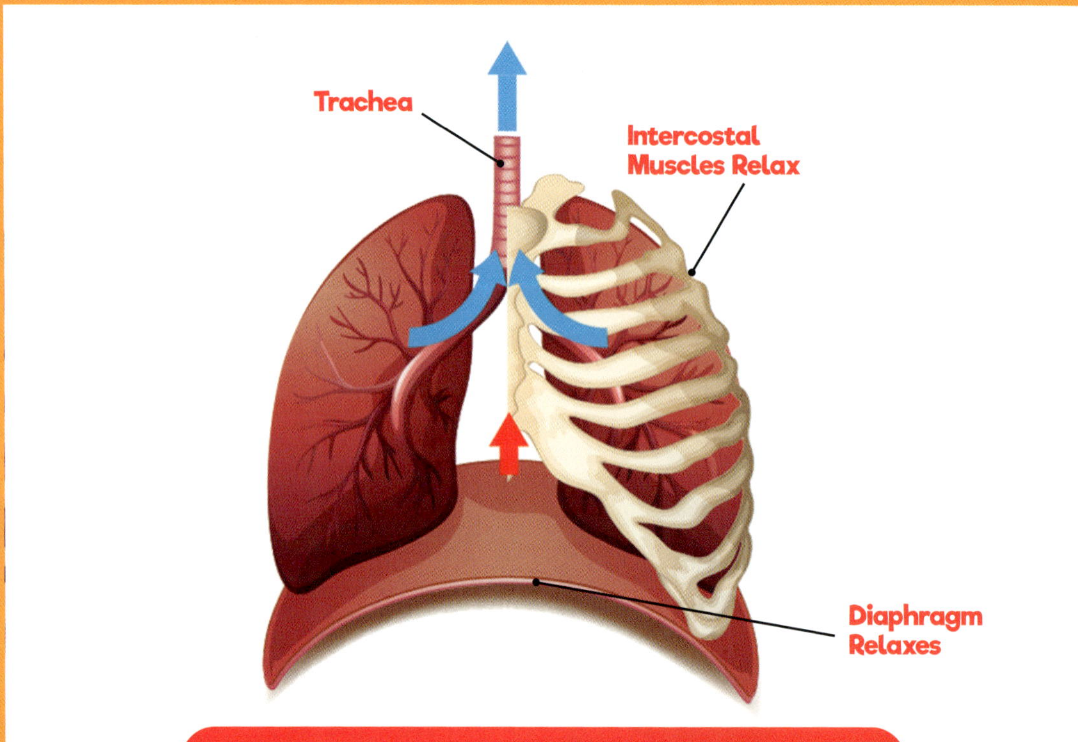

The diaphragm relaxes when we breathe out.

When it's cold outside, we can see our breath as we exhale. This is the water in our breath cooling to form water vapor.

The red blood cells carry the waste gas back to the lungs. They travel through the capillaries and send the waste gas into the alveoli. The diaphragm muscle then pushes the waste gas through the bronchioles, out the bronchi, up the windpipe, and finally out the nose or mouth.

Air that is exhaled is warm because it picks up heat as it travels through the body. This is why you can see your breath as water vapor when it's cold outside.

Consider This

Taking long, deep breaths can help relax the body. Why do you think this is?

Illnesses of the Lungs

There are many illnesses that affect the lungs. Some people have a breathing problem called asthma. Their bronchioles sometimes tighten so that air cannot get into the alveoli. It is one of the most common long-term health problems in children.

People with asthma may use an inhaler. This is a tool that gives the lungs a dose of a medicine that opens the bronchioles, making breathing easier again.

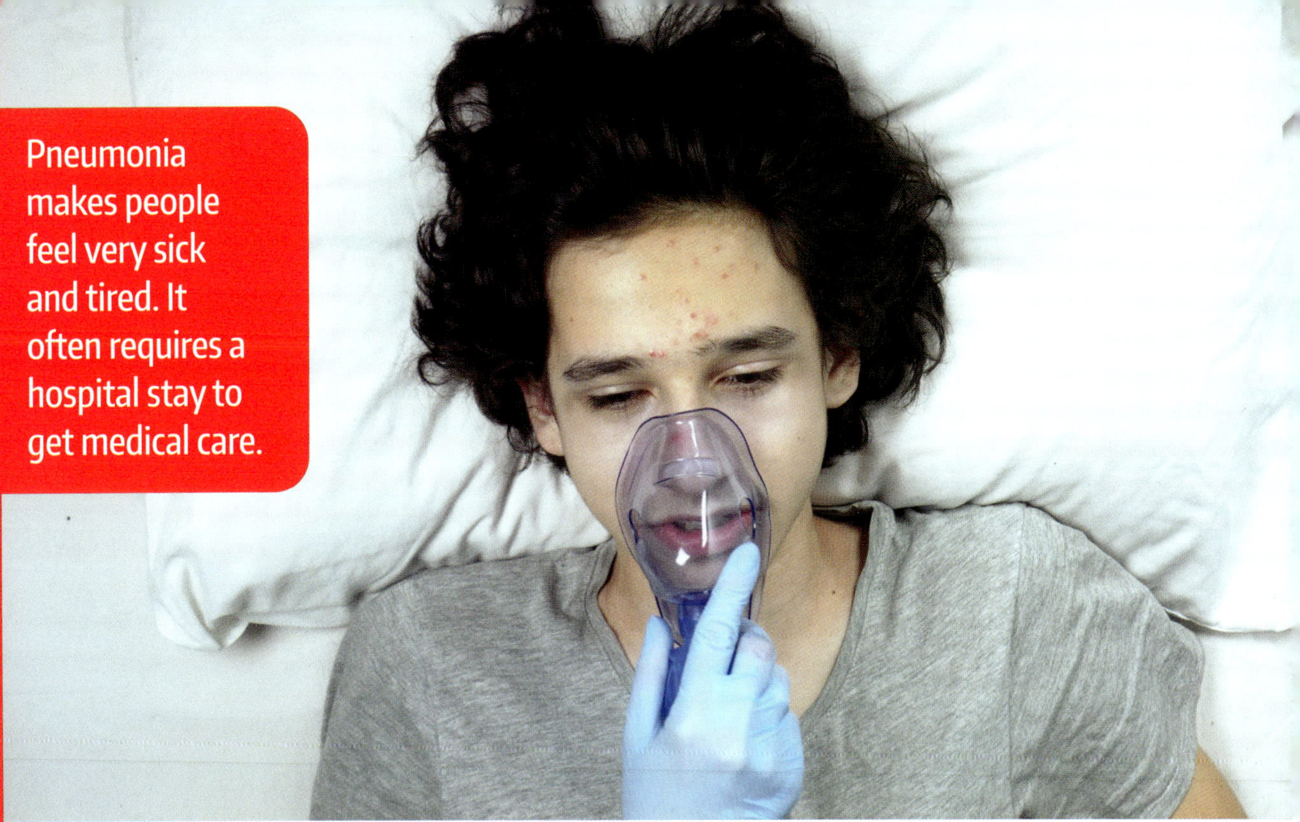

Pneumonia makes people feel very sick and tired. It often requires a hospital stay to get medical care.

Another sickness is bronchitis, which is when the bronchi are sore or damaged. Bronchitis makes it hard for air to get to the alveoli. People with bronchitis cannot breathe easily, cough a lot, and feel pain in their chests. An illness called pneumonia causes the lungs to swell. The lungs have a hard time absorbing the oxygen in the air.

Consider This

It's possible to live with one lung. This might happen due to illnesses or accidents. How might this change the way a person lives?

Smoking and Vaping

Smoking is very bad for the lungs. Smoke damages cilia so that they cannot keep dirt and other harmful things out of the lungs. Smoke also breaks down the walls of the alveoli.

In many ways, the lungs are strong. However, any kind of smoke can easily damage them. Firefighters often warn that more people die from inhaling the smoke of a fire than from burn injuries.

Vaping liquids can contain nicotine (an addictive substance found in cigarettes as well), heavy metals such a lead, and cancer-causing chemicals.

Over time, inhaling cigarette smoke causes the lungs to become so damaged that they are unable to heal. Emphysema or lung cancer can result. Both are deadly diseases of the lungs. People who inhale smoke in the air from other people's smoking can also get these diseases.

Vaping is similar to smoking but it is a much newer activity. Scientists are still studying the affects of vaping on the human body and lungs.

Consider This

Smoking and vaping don't just affect the lungs. Thay can have negative affects on other parts of the body as well. Can you think of an example?

Healthy Lungs, Healthy Body

Think of all the ways your lungs help you during a day. They do more than help you breathe, talk, or shout. One could argue that the lungs help the entire body do everything that it does. That is why it is important to keep your lungs healthy!

You can help your lungs stay healthy by not smoking and vaping. Other pollution in the air, such as exhaust from cars, can also hurt the lungs.

You can keep your lungs healthy by avoiding smoking, and by getting plenty of exercise.

Getting exercise can be as simple as playing basketball with friends.

Exercising is a good way to keep your lungs strong and healthy. It is good for your whole body, too. When you exercise, your body has to do more than it usually does. For example, running requires your legs to move faster. Your body cells need more oxygen to do that extra work. You must breathe harder and faster to get oxygen to your cells. That makes your lungs work harder, which helps make your whole body stronger and healthier.

Consider This

What are your favorite healthy ways to raise your rate of breathing and get your heart beating faster?

Glossary

alveoli Tiny air sacs at the end of each bronchiole.
asthma A physical condition that makes it hard for the lungs to breathe.
blood vessels Small tubes that carry blood to different parts of the body.
bronchi The two tubes that lead from the windpipe into each lung; a single tube is called a bronchus.
bronchioles Small branches of the bronchus.
bronchitis An illness in which the bronchial tubes become sore or damaged and coughing results.
capillaries The smallest tubes that carry blood through the body.
carbon dioxide A colorless waste gas that is made when people breathe.
cell A very small, basic unit of all living things.
diaphragm A dome-shaped sheet of muscle under the lungs that is used for breathing.
emphysema A serious disease of the lungs that is often caused by smoking.
mucus A thick liquid that is made in some parts of the body for protection.
organ A part of the body that has a specific job to do.
oxygen A gas found in air that is needed by most life-forms on Earth to live.
pneumonia A serious illness that affects the lungs and makes it hard to breathe.
rib cage The bony frame that protects organs that are located in the chest.
tissue A family of cells that live close together and work together to do the same job.
vocal cords Thin pieces of folded tissue that help the voice make sounds.
voice box Part of the throat that holds the vocal cords; also called the larynx.
windpipe A long tube from the neck to the chest that carries air in and out of the lungs; also called the trachea.

For More Information

Books

Solway, Andrew. *Heart and Lungs*. Buffalo, NY: Gareth Stevens Publishing, 2025.

Xiong, Keng. *What Is the Respiratory System?* Parker, CO: The Child's World, 2025.

Websites

Breathing and the Respiratory System

www.ducksters.com/science/breathing.php

Read more about the respiratory system here and find links to other related topics.

Your Lungs and Respiratory System

kidshealth.org/en/kids/lungs.html

Kids Health has an informative collection of facts about the respiratory system.

Index

A
alveoli, 18, 19, 20, 21, 23, 24, 25, 26
asthma, 24

B
brain, 5
bronchial tubes, 8, 9, 16, 23, 25
bronchioles, 16, 17, 18, 23, 24
bronchitis, 25

C
capillaries, 18, 20, 23
carbon dioxide, 10, 11, 22
cilia, 14, 15, 26

D
diaphragm, 6, 7, 8, 20, 21, 22, 23

E
esophagus, 9
exercise, 28, 29

H
heart, 5, 6, 11, 16, 20

I
illness, 14, 15, 24, 25

M
mouth, 8, 9, 12, 13, 20, 23
mucus, 14, 15

O
oxygen, 4, 10, 11, 18, 19, 20, 22, 25, 29

R
rib cage, 5, 6, 20

S
smoking and vaping, 26, 27, 28

V
vocal cords, 12, 13
voice box, 8, 9, 12

W
windpipe, 8, 9, 12, 16, 20, 23